Particular Odyssey: In Search

Particular Odyssey: In Search

Meg Withers

Original cover art by J. Isaac Evans
Book design, Marsea Guevara
Published by Prickly Pear Publishing
Santa Fe, New Mexico

Printed in the United States of America

For information, contact Prickly Pear Publishing, Santa Fe, NM
www.pricklypearpublishing.com

Library of Congress Cataloging-in-Publication Data

Names: Withers, Meg, author Title: Particular Odyssey: In Search
Description: Santa Fe : Prickly Pear Publishing [2021]

Particular Odyssey: In Search / Meg Withers 1st Prickly Pear Publishing ed.p.cm

Poetry
ISBN: 978-1-889568-05-8

10 9 8 7 6 5 4 3 2 1

Dedication

To those who have persisted in being helpmates and hanai-style family, and who have given me time and love. This is also to say that many on this list aren't aware of being teachers: they're just fine human beings, hence teachers. First, let me thank ka'Āina, the Hawaiian islands, for the inspiring wisdom of earth. To Arthur Sze, tamer of baroque opera drama queen modifiers and excessive other verbiage, and to El Maestro, Francisco X. Alarcón, who saw me at a workshop in 1994, sat down to talk, and became a mentor and friend for life. To AD Miller, one of the divine souls who also mentored me into poetry. He was always there. To Truong Tran, eater of words, garlic, and phở, who helped me to break all the rules I could get away with, being here for the broken parts that transformed themselves into some crazy form of grace. To Odilia Galván Rodríguez, hermana, compatriot, poet/activist, and extraordinary human being – no nonsense creative genius, and creator of the Hawk, who is also responsible for mentoring me through his mother. To Zaid Shlah, brother, writer-friend and fellow poet, who encouraged me in the magic of storytelling. And, let me not forget the great master grokker of the persistent oddities and similarities of humans: Italo Calvino. Many names should be here – probably all of them, as this has been an extended odyssey, and I have learned some lessons. Mahalonuiloa!

Particular Odyssey: In Search

...everything is everything...pass the joint

— Donny Hathaway, "The Ghetto"

There is sometimes confusion surrounding the subject of *mass* in relativity. This is because there are two separate uses of the term. Sometimes people say "mass" when they mean "relativistic mass", *mr* but at other times they say "mass" when they mean "invariant mass", *m0*. These two meanings are not the same. The invariant mass of a particle is independent of its velocity *v*, whereas relativistic mass increases with velocity and tends to infinity as the velocity approaches the speed of light *c*.

— Taylor and Wheeler, *Space-Time Physics*

...i paralleli, i paralleli...

— Umberto Ecco's recollection of Italo Calvino's last words

An entangled pair of objects—photons, electrons, whatever—can propagate and evolve together as a unit, even while moving off in different directions, but only until one member of the entangled pair is observed. At that instant the entangled system is destroyed and a single eigenstate is detected.

— Stephen G. Benka, "The Entangled Dance of Physics"

Table of Contents

Particular Odyssey: In Search

Into/Out/.....

Recent photos of cellular structure bear a remarkable resemblance to pictures of the Universe. Both are the basis for life (what we know of it). Imagine that everyone lives in/is their own universe – all at once. If this sounds like too many hits of psilocybin or too much LSD; it was a good trip. The truth is in Donny Hathaway's lyrics, "The Ghetto" (1970): "Everything is everything."

Poetry makes connections between ideas, any living being (and some non-living), concepts, universes, languages. It's a great big neighborhood, which is not to say that all is well here. Poetry is one of the greatest forms of reconciliation and has as much conductivity as the most sophisticated electronic instrument. Most artists feel like this about their medium. Take the word medium for instance, which is a whole other universe.

It's a Hokey Pokey world no doubt about it. We come here, then leave, and maybe come here again, or not. It's actually none of our business what happens, as long as we make the kinds of connections that attempt to perform reconciliation and conductivity. The lefthand in, then the left hand out. We're turned about in ways we could never imagine – as long as we don't interfere with universal rules. And, the rulebook is huge and lovely, magical with curves and immense trajectorial possibilities.

The Sciences and Maths are the gravity that pull us into this version of reality. Information is a renewable passport to the ever-changing; changing the way the universe looks and acts – or so it appears to us. It was probably already that way to begin with: we just didn't know. Now we have a new piece of this gigantic puzzle. We fly on through the immensity.

Language is the same; continually remaking itself and the meaning of existence. A group of folks moves into the new neighborhood, and language is hatched from the words of travelers – reconciliation and conductivity are given an opportunity to create themselves. Every time something new is discovered about how the universcell looks or functions, we open the door into what we may think is the Answer to It All. Mostly, we are faced with another circuitous hallway with countless unopened doors. This is the real magic oflanguage, where all the voices belong. This book is a vision of a particular odyssey toward connectedness.

"Pass the joint…."

OVERTURE:

if on a splinter

nigh *a positron*

strong nuclear force holds

this peach

the bee a bone

a positron

not just humming sensation

that hammering noise

anti-matter

light *exists*

let us continue to rob bees

naked

singularity

see
through everything
nano possible

force
jumps
the *cat*

fluttering
curtains

orbital
magnet

we are sure

photons reach
cannot live
without

cloud
chambers

"everything
that rises
must converge"

blue
the last color
we see
ascend -
ing

plumage   snowy egret

mirrors   reeds   green pond

replicates   mallard

iridescence

calla's

receptacle

echo chamber

koan

auricle/ventricle

the speed at which stars rotate
around your galaxy
reflects the amount
of *dark matter*
surrounding you

birth   death   phonon   particles

occur

space   0.11mm

we cannot sense

chaos as order

in cooling water to form ice crystals
the *perfect* symmetry of *water*

is lost

tra   jec   tory

purple blur poised

edge

arrays   lambda

travel

outside   *art*

Malbolge

99.99%

mc

□’d

single

excited string

FIRE

think   membrane

believe   we stand

we merely   jump off

quark bounce   muon   flight

dark matter   dwarf star

implode

baroque   *crimson*

reverse   Doppler

echo   burning

render   to   null

void

speed of   source

divided   Bach

wing   feather   *hawk*

Space and time cannot be separated

radiation

matter

world

made

flesh *dwelt*

among

spake

existence

caught

the path

single

spiral

electron

song

substance

escapes

black force

hawking lost his

baseball book

≈

iridescent motion

butterfly wings

thought *threshold*

tornado

≡

anti-matter

subsumes

what *is* not
ought but 0

whirl
*star*lings

tell tale *spots* cosmos

shower

feathered

yellow-beaked

birth

ruffled peonies

leaning

whisper *need*

steep slope

slow coils vortex

finding home

luminous

city

rosemary is *known*

to heal *memory*

i was

the

universe

and

born

∞

AIR

without fear

wind or

vertigo sky

tends *exotic*

turquoise tremolo

blush sunrise pions

soprano rain

speaks

certain

i've forgotten

nothing *superconductive*

full pale

late

winter

precocity of *women*

cockade

swerve

phalarope *dancing*

sound listens

up

cathedral *twilight*

solar *flare*

blue saxophone

without precedent meson fires

scorched *marigold* prairie

bruised sky

muon sonata

dark *Mozart*

madder

eagle aerie

lift *euphoria*

blooms

cosmic dance dress

gamma *rays*

cerise veil

rose

*smolder*ing

copper

swirl

we saw in ga*lax*ies
crystals resembling crushed *glass*.

winter freeze

lake of slow fish

flash ing

looks gather

silver

neutrinos

escape *slower*

cold wind shadow

rainbow mouth

bubbles

move on

spring

b-flat

=

ripe water*melon*

=

hammerklavier sonata

Beethoven blind

inner vibrato

tapping summer

melon rind

aura

∂

pressure ripples sound waves

human speech

billion spans traveling

dark matter sing

concentric *riffs*

network lines

enlace

30,000 *light* years

sway

abso*lute* darkness
does not exist
unless imagined

WATER

lagune *proteus*

element

sapid

mechanical

fluid

art

motion

long

cogent

drop

[gap]

closed

hand

opens

palm light

flip

nuclear catalyst

gold yolk

solar

chronos

pan-optic laser

coherent

steers *vertical*

narrative

net

life

1100100100110001101000101100

is/not

null/one

path lines

oblivion

cerebrum

soul

29 grams

what is *lost*

intersects

aurora

black pearl

swivel

nacre creation

sun heat

rotation

*pulse*d particles

revolve mesons

leptons

excitation

flutter doves

tulle *clouds*

toe-shoes

universe

en pointe

flamboyance

drift

flight

muscular

undertow

exaltations

honeyeaters rare copper

charm quark

hourly evanesce

filaments ignite

contiguous *bees*

magic

hexagonal drone source

radiance

does not fail

nothing in the universe
lacks vibration

marine

bruise blue *fails*

contralto

parlance deep

/hydrogen
/hydrogen

/oxygen

fluoresce everything

play velvet contrapuntal

kelp

memory patterns

DNA dance

float *riffle* strand

trail fire nexus

pulse

seagrass

starfish

anemone *weave*

.

strange *surrender*

molecules circle return

parallax concave

Hayden

connects rush

leaves

parachute

ray affection

calm

amplitude

sways watermusik

coral

includes

cerulean

folds

black holes
are actually dark green

creations

substrata

≈

certain

selky

creatured *supple* moon

dense undergrowth

soul silicate

limpid

formless

cupped

finite prerogatives shaped liquid

possibility

enformed

fluid

hand *fills*

shadow's flight

glistens

stasis

eight-legged eye giant

escaping flagellant

beak-mouthed sinuous

svelte *cursive*

manifold empty grave

silver opened

slippers *circle*

infinite

waterborne

yes

is left

the *closer* the answer
the larger the *question*

dream

slip

turn

together

profound

voluminous

quiescent

breath

breathe

down

O
2

r i s e

reflective *flow*

al-‘Adhara maidens

al-Khiba al-Firq

tent flock

crepuscular Orion

ancestor

blink

creatures

spangled *wise*

encoded *spiral*

lumen pheromone

not one but many

helios *helical* helix

kaleidoscope

map *feathered*

message

EARTH

calla graph *das Liede*

Mahler baritone

warp/weft schist

engage

lowing *dwell*

striated

vulcan wields

strike/slip *fulcrum*

sub molten

a'a

laupahoehoe sex

karmic tantrum

magmatic *arc*

thrust-fault rapid

subduction lavaic *moon*

milk *lens*

pro

ta

zoa

wriggle

writhe

sperm

insouciant

hipshot

spun

life

trance

heat

swim

split

wind

now

form

breath

each spring
wisteria

gilded *nucleon*

ion quest

delft irid

myriad tracks *unity*

telluric

radiation *flicker*

what *story*

there

awaits

euphony

is

ever…

even at z zero mass
nothing ever dies

CODA:

if on a splinter

a positron

light exists

a cat

is

cloud

ascending

calla's receptacle

koan

dark matter

chaos as

order

perfect

water

is lost

or rays

of art

99.99%

mc

□’d

think

crimson

echo

speed wing

hawk

matter dwelt

caught

song

substance

iridescent threshold

as ought

star spots

birth

need

slow finding

known

memory

born

exotic

superconductive

sky

water

certain

full women

swerve

dancing

twilight

flare

marigold

Mozart madder

lift

euphoria

ray

smolder

lax

glass

winter

lake

slower

rainbow

spring

ripe melon

aura

pressure

dark riffs

network lines

enlace

light

lute proteus

sapid

motion

opens

solar

coherent

vertical

path

lost nacre pulse

revolve

flutter

tulle clouds

en pointe

drift

hourly

bees

magic

nothing

fails

contralto parlance

sways

riffle

pulse

weave

surrender

parachute

includes folds

certain

supple

cupped

fill

shadow

glistens

escaping

cursive

circle

yes

closer

question

is left

voluminous

breath

rise

flow

blink

creaturely wise

spiral

helical

feathered

message

das Liede

dwells

fulcrum

a’a

laupahoehoe arc

moon

lens

life

wind

now

nucleon

unity

flicker

ion story

awaits

its

ever…

Gloss Poetica

Mathematical Symbols

∞	infinity
$\equiv$	always equals
$\approx$	is approximately equal to
$\int$	integral
$\neq$	does not equal
∂	partial differential

Arabic

al-`Adhara: the Maiden – a September constellation in Canis Majoris

al-Khiba: the Tent – a May constellation near Hydra - also known as "The Raven"

al-Firq: the Flock – an October constellation, Cepheus, who was the father of Adromeda

German

das Liede: The song (from Mahler's *das Liede von der Erde – the Song of the Earth*)

Celtic

selky: a seal which magically transforms itself to a beautiful woman

Hawaiian

a'a: A type of cooled lava that is jagged and produces shard-like edges.

laupahoehoe
or Pahoehoe: (also an adapted Geological term): A basaltic lava flow with a glassy, smooth, and undulating, or ropy, surface.

Atomic and Sub-atomic Particles

Muons: A type of heavy electron-like particle, millions of which rain down each second from space.

Gamma Rays: Photons, particularly high energy photons.

Bosons: Intrinsic angular momentum (integer spin in units of Planck's constant).

Hadrons: (Heavy) mesons and baryons which are made of quarks.

Photons: A boson - a particle that reacts according to Einstein's theories of particle activity.

Mesons: Particles whose basic structure is a quark and an anti-quark.

Pions: Pi mesons - lightest mesons.

Leptons:	Tau, and three types of neutrinos, include muons, electrons, two types of neutrinos, and respective antiparticles, including positrons – the derivation of the word is "light" from its characteristic of being lighter than the lightest meson.
Quarks:	Any of a group of six elementary particles having electric charges of a magnitude one-third or two-thirds that of the electron, decay by weak nuclear force. Six "flavors" of quarks include: strange, charm, up, down, top, and bottom. The term was coined from James Joyce's Finnegan's Wake.
Positron:	The opposite of an electron – a lepton.
Electron:	The negatively charged particles that form the outer valence structure of any atom.
Kaon:	A subgroup of unstable mesons that consist of a strange quark and a light anti-quark produced as a result of a high-energy particle collision.
Lambda:	A baryon particle made with a strangeness quantum number 1.
Neutrinos:	Neutral particle with very tiny mass, produced in weak interaction decay.
Nucleon:	A proton and neutron.
Phonons:	Tiny vibrations of a crystal lattice.
Fermions:	Particles characterized by spin in odd half integer quantum units.

Ion: An atom either lacking or containing additional electrons.

Other Physics and Astronomic Terminology

z zero mass: The mass of the z zero – particle that mediates neutral weak interactions.

Strong Nuclear Force: the force that holds nuclei together.

Dark Matter: Physical objects or particles that emit little or no detectable radiation of their own and are postulated to exist because of unexplained gravitational forces observed on other astronomical objects.

Dwarf Star: A star which is on the verge of collapsing into a black hole. Doppler Shift: The way a sound shifts in tone coming and going from a point.

Light Year: The distance light travels in a year – actually a measure of distance.

Coherent Light: A beam of light in which the photons are in phase with one another.

Exclusion Principle: The principle which states that you cannot have two Fermions in the same state in the same place at the same time. This principle gives us the entire structure of chemistry.

Bose-Einstein Condensate: A condition in which innumerable bosons can exist in a coherent state.

Uncertainty Principle. A counterintuitive mainstay of modern physics, this asserts that in the subatomic realm of waves and particles, to observe a phenomenon is to alter it.

Poincaré's conjecture - A rabbit is the same as a sphere - Poincaré's asserts that if any loop in a certain kind of three-dimensional space can be shrunk to a point without ripping or tearing either the loop or the space, the space is equivalent to a sphere.

Geological Terminology

Strike-slip: A fault whose relative displacement is purely horizontal

Magmatic arc: The trajectory at which molten rock material travels upward. Magma that reaches the surface is referred to as lava.

Thrust-fault: A dip-slip fault in which the upper block above the fault plane moves up and over the lower block, so that older strata are placed over younger.

Rapid subduction: A rapidly dipping planar zone descending away from a trench and defined by high seismicity, interpreted as the shear zone between a sinking oceanic plate and an overriding plate.

Schist: A metamorphic rock characterized by strong foliation or schistosity.

Foliation: Any planar set of minerals or banding of mineral concentrations including cleavage, found in a metamorphic rock.

About The Author

When people ask Meg Withers where she's from, she never knows what to say. She was a Navy kid, who traveled from the time she was born, right after WWII. She lived all over the U.S., including ten years in Hawai'i, and a year in England. Someone would say: "Hey, let's go to Asheville," and the next month she moved again. Along the way she met hundreds of lovely human beings, wrote poetry the whole time, connected and reconnected like a vagrant planet, had family, who gave up on her wandering, shrugging and letting her go, or went looking for her just to see what the heck she was up to now. Her cousin Lidia once told her: "You're the busiest little kid I've ever known." Her nonna constantly asked: "La Margarita. Vieni qui? Dove stai andando adesso?" Come here. Where are you going now? Who knows? She doesn't. She's the author of three books of poetry, Must Be Present to Win, A Communion of Saints, and the one in your hands, the poetry editor and contributor to a poetry anthology about women, Shadowed: Unheard Voices, and is currently working on a memoir of sorts, called So, We Get To Talkin'. It begins after she is dead. One of these years she'll retire, but she doesn't have a fishing pole. Maybe she'll put up a yurt. Easier to travel.

Acknowledgements

This piece owes its conception and implementation to Italo Calvino, and to his love of the parallels natural to existence. The lines of the titles of the named chapters of *if on a winter's night a traveler,* are modified and embedded in the lines of this poem, and for the use of them I am grateful.

A great deal of gratitude goes to Professor Sarah Church, astrophysicist and cosmologist, and head of the Kavli Institute at Stanford University, who has given kind regard and support for this project.

I am also grateful to Dr. Helen Quinn, of Stanford Linear Accelerator Laboratory, for the time she spent explaining particle physics terms and relationships to me one day.

The checks for astrophysics and particle physics prior to the publication of the entire manuscript were checked by Drs. Church and Quinn sometime around 2011.

Made in United States
Orlando, FL
05 November 2026

90241458R00048